Prentice Hall Science

AF696771

Laboratory Manual

MATTER

Building Block of the Universe

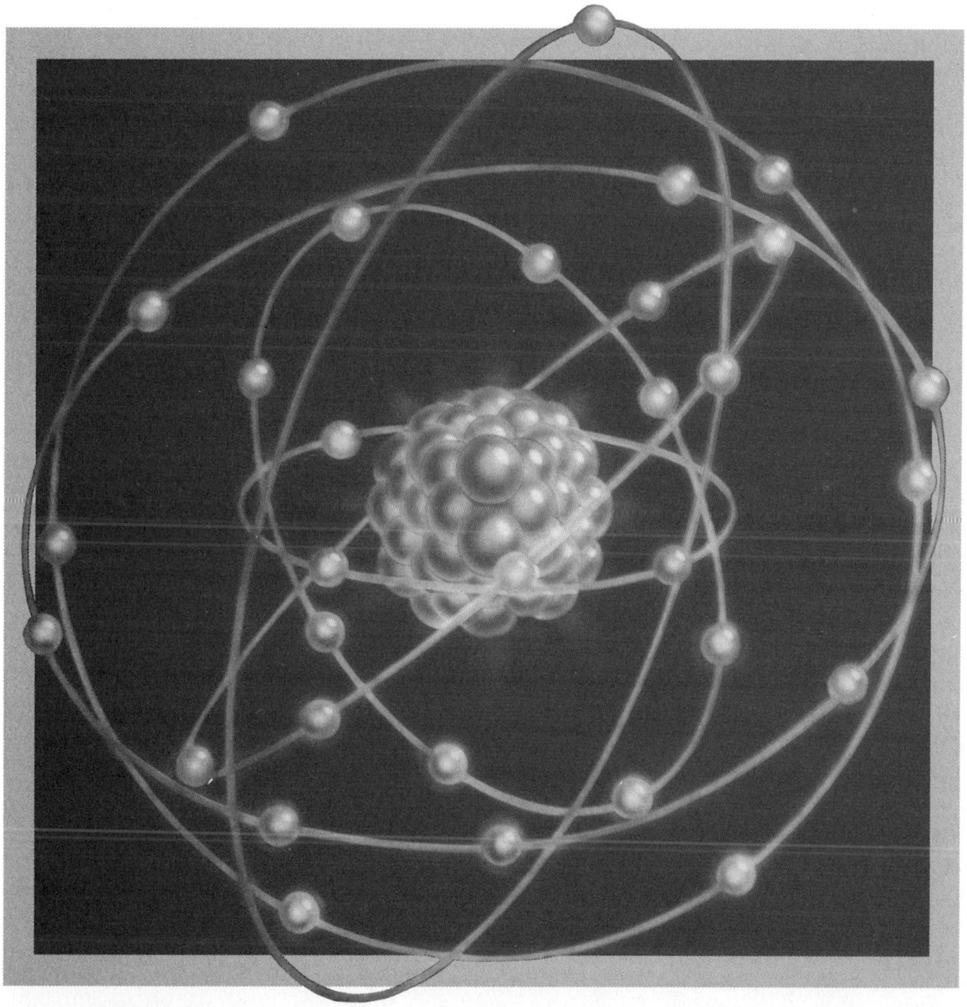

Prentice Hall
Englewood Cliffs, New Jersey
Needham, Massachusetts

Laboratory Manual

PRENTICE HALL SCIENCE
Matter
Building Block of the Universe

ISBN 0-13-986134-3

5 6 00 99

Prentice Hall

Englewood Cliffs, New Jersey 07632

Contents

Safety Symbols

All the investigations in this *Laboratory Manual* have been designed with safety in mind. If you follow the instructions, you should have a safe and interesting year in the laboratory. Before beginning any investigation, make sure you read the safety rules that follow.

The eight safety symbols below appear next to certain steps in some of the investigations in this *Laboratory Manual*. The symbols alert you to the need for special safety precautions. The description of each symbol below tells you which precautions to take whenever you see the symbol in an investigation.

Glassware Safety

1. Whenever you see this symbol, you will know that you are working with glassware that can easily be broken. Take particular care to handle such glassware safely. And never use broken or chipped glassware.
2. Never heat glassware that is not thoroughly dry. Never pick up any glassware unless you are sure it is not hot. If it is hot, use heat-resistant gloves.
3. Always clean glassware thoroughly before putting it away.

Fire Safety

1. Whenever you see this symbol, you will know that you are working with fire. Never use any source of fire without wearing safety goggles.
2. Never heat anything—particularly chemicals—unless instructed to do so.
3. Never heat anything in a closed container.
4. Never reach across a flame.
5. Always use a clamp, tongs, or heat-resistant gloves to handle hot objects.
6. Always maintain a clean work area, particularly when using a flame.

Heat Safety

Whenever you see this symbol, you will know that you should put on heat-resistant gloves to avoid burning your hands.

Chemical Safety

1. Whenever you see this symbol, you will know that you are working with chemicals that could be hazardous.
2. Never smell any chemical directly from its container. Always use your hand to waft some of the odors from the top of the container toward your nose—and only when instructed to do so.
3. Never mix chemicals unless instructed to do so.
4. Never touch or taste any chemical unless instructed to do so.
5. Keep all lids closed when chemicals are not in use. Dispose of all chemicals as instructed by your teacher.
6. Immediately rinse with water any chemicals, particularly acids, that get on your skin and clothes. Then notify your teacher.

Eye and Face Safety

1. Whenever you see this symbol, you will know that you are performing an experiment in which you must take precautions to protect your eyes and face by wearing safety goggles.
2. When you are heating a test tube or bottle, always point it away from you and others. Chemicals can splash or boil out of a heated test tube.

Sharp Instrument Safety

1. Whenever you see this symbol, you will know that you are working with a sharp instrument.
2. Always use single-edged razors; double-edged razors are too dangerous.
3. Handle any sharp instrument with extreme care. Never cut any material toward you; always cut away from you.
4. Immediately notify your teacher if your skin is cut.

Electrical Safety

1. Whenever you see this symbol, you will know that you are using electricity in the laboratory.
2. Never use long extension cords to plug in any electrical device. Do not plug too many appliances into one socket or you may overload the socket and cause a fire.
3. Never touch an electrical appliance or outlet with wet hands.

Animal Safety

1. Whenever you see this symbol, you will know that you are working with live animals.
2. Do not cause pain, discomfort, or injury to an animal.
3. Follow your teacher's directions when handling animals. Wash your hands thoroughly after handling animals or their cages.

Science Safety Rules

One of the first things a scientist learns is that working in the laboratory can be an exciting experience. But the laboratory can also be quite dangerous if proper safety rules are not followed at all times. To prepare yourself for a safe year in the laboratory, read over the following safety rules. Then read them a second time. Make sure you understand each rule. If you do not, ask your teacher to explain any rules you are unsure of.

Dress Code

1. Many materials in the laboratory can cause eye injury. To protect yourself from possible injury, wear safety goggles whenever you are working with chemicals, burners, or any substance that might get into your eyes. Never wear contact lenses in the laboratory.
2. Wear a laboratory apron or coat whenever you are working with chemicals or heated substances.
3. Tie back long hair to keep it away from any chemicals, burners, and candles, or other laboratory equipment.
4. Remove or tie back any article of clothing or jewelry that can hang down and touch chemicals and flames.

General Safety Rules

5. Read all directions for an experiment several times. Follow the directions exactly as they are written. If you are in doubt about any part of the experiment, ask your teacher for assistance.
6. Never perform activities that are not authorized by your teacher. Obtain permission before "experimenting" on your own.
7. Never handle any equipment unless you have specific permission.
8. Take extreme care not to spill any material in the laboratory. If a spill occurs, immediately ask your teacher about the proper cleanup procedure. Never simply pour chemicals or other substances into the sink or trash container.
9. Never eat in the laboratory.
10. Wash your hands before and after each experiment.

First Aid

11. Immediately report all accidents, no matter how minor, to your teacher.
12. Learn what to do in case of specific accidents, such as getting acid in your eyes or on your skin. (Rinse acids from your body with lots of water.)
13. Become aware of the location of the first-aid kit. But your teacher should administer any required first aid due to injury. Or your teacher may send you to the school nurse or call a physician.
14. Know where and how to report an accident or fire. Find out the location of the fire extinguisher, phone, and fire alarm. Keep a list of important phone numbers—such as the fire department and the school nurse—near the phone. Immediately report any fires to your teacher.

Heating and Fire Safety

15. Again, never use a heat source, such as a candle or a burner, without wearing safety goggles.
16. Never heat a chemical you are not instructed to heat. A chemical that is harmless when cool may be dangerous when heated.
17. Maintain a clean work area and keep all materials away from flames.
18. Never reach across a flame.
19. Make sure you know how to light a Bunsen burner. (Your teacher will demonstrate the proper procedure for lighting a burner.) If the flame leaps out of a burner toward you, immediately turn off the gas. Do not touch the burner. It may be hot. And never leave a lighted burner unattended!
20. When heating a test tube or bottle, always point it away from you and others. Chemicals can splash or boil out of a heated test tube.

21. Never heat a liquid in a closed container. The expanding gases produced may blow the container apart, injuring you or others.
22. Before picking up a container that has been heated, first hold the back of your hand near it. If you can feel the heat on the back of your hand, the container may be too hot to handle. Use a clamp or tongs when handling hot containers.

Using Chemicals Safely

23. Never mix chemicals for the "fun of it." You might produce a dangerous, possibly explosive substance.
24. Never touch, taste, or smell a chemical unless you are instructed by your teacher to do so. Many chemicals are poisonous. If you are instructed to note the fumes in an experiment, gently wave your hand over the opening of a container and direct the fumes toward your nose. Do not inhale the fumes directly from the container.
25. Use only those chemicals needed in the activity. Keep all lids closed when a chemical is not being used. Notify your teacher whenever chemicals are spilled.
26. Dispose of all chemicals as instructed by your teacher. To avoid contamination, never return chemicals to their original containers.
27. Be extra careful when working with acids or bases. Pour such chemicals over the sink, not over your workbench.
28. When diluting an acid, pour the acid into water. Never pour water into the acid.
29. Immediately rinse with water any acids that get on your skin or clothing. Then notify your teacher of any acid spill.

Using Glassware Safely

30. Never force glass tubing into a rubber stopper. A turning motion and lubricant will be helpful when inserting glass tubing into rubber stoppers or rubber tubing. Your teacher will demonstrate the proper way to insert glass tubing.
31. Never heat glassware that is not thoroughly dry. Use a wire screen to protect glassware from any flame.
32. Keep in mind that hot glassware will not appear hot. Never pick up glassware without first checking to see if it is hot. See #22.
33. If you are instructed to cut glass tubing, fire-polish the ends immediately to remove sharp edges.
34. Never use broken or chipped glassware. If glassware breaks, notify your teacher and dispose of the glassware in the proper trash container.
35. Never eat or drink from laboratory glassware.
36. Thoroughly clean glassware before putting it away.

Using Sharp Instruments

37. Handle scalpels or razor blades with extreme care. Never cut material toward you; cut away from you.
38. Immediately notify your teacher if you cut your skin when working in the laboratory.

Animal Safety

39. No experiments that cause pain, discomfort, or harm to mammals, birds, reptiles, fish, and amphibians should be done in the classroom or at home.
40. Animals should be handled only if necessary. If an animal is excited or frightened, pregnant, feeding, or with its young, special handling is required.
41. Your teacher will instruct you as to how to handle each animal species that may be brought into the classroom.
42. Clean your hands thoroughly after handling animals or the cage containing animals.

End-of-Experiment Rules

43. After an experiment has been completed, clean up your work area and return all equipment to its proper place.
44. Wash your hands after every experiment.
45. Turn off all candles and burners before leaving the laboratory. Check that the gas line leading to the burner is off as well.

Name ______________________ Class ______________ Date ____________

Laboratory Investigation

General Properties of Matter **1**

Determining Density

Background Information

Which is heavier, a kilogram of lead or a kilogram of feathers? This is an old question with a simple answer. Since both objects have a mass of 1 kg, they are equally heavy. Each feather, however, has less mass than each piece of lead. So a greater number of feathers is needed to make a kilogram. And the more feathers there are, the more space they take up. The space an object occupies is its volume. So a kilogram of feathers has a greater volume than a kilogram of lead. The mass of a certain volume of feathers is different from the mass of the same volume of lead.

The mass of a specific volume of an object is called density. Density can be expressed mathematically as:

$$\text{Density} = \frac{\text{Mass}}{\text{Volume}}$$

Density is a property of all objects. In this investigation you will learn about this important characteristic.

Problem

How can the density of an object be determined?

Materials *(per group)*

triple-beam balance
metric ruler
100-mL graduated cylinder
2 rectangular solids of the same material but of different dimensions, such as a bar and a cube of iron, copper, or aluminum
1 large piece of modeling clay

Procedure

Part A Density of Rectangular Solids

1. Using the balance, determine the mass of each solid. Follow the directions given by your teacher for the proper use of the balance. Read the mass of each solid to the nearest 0.1 g. Record your data in the appropriate column of Data Table 1.

2. Determine the volume of each solid using the metric ruler to measure length, width, and height. Read these values to the nearest 0.1 cm. Now use the formula for volume $V = L \times W \times H$, where L = length, W = width, and H = height, to calculate the volume of each solid. Add these data to Data Table 1.

3. Calculate the density of each sample, using the formula Density = Mass/Volume ($D = M/V$), and record your results in Data Table 1.

Part B Density of Irregular Solids

1. Separate the large piece of modeling clay into five pieces of different sizes. Make sure that the mass of the smallest sample is no less than 5 g.

2. Using the balance, determine the mass of each sample to the nearest 0.1 g. Record your data in the appropriate column of Data Table 2.

3. The volume of an irregular solid cannot be determined in the same way as a rectangular solid, since length, width, and height cannot be accurately measured. However, the volume of an irregular solid can be measured by the displacement of water.

4. Place some water in the graduated cylinder and read its volume to the nearest 0.1 mL. Carefully add a sample of clay, making sure you do not let any water splash out of the graduated cylinder. Read the new volume to the nearest 0.1 mL. Subtract the original volume from the new volume. This volume of displaced water is equal to the volume of the clay sample. Record this value in Data Table 2.

5. Repeat the procedure for the other four samples of clay. Record your data in Data Table 2.

6. Using the mass and volume values in Data Table 2, construct a graph that illustrates the relationship between mass and volume of each sample. Use the vertical axis for mass and the horizontal axis for volume.

7. Connect the five points on the graph and extend the line to the zero point. The line should be a straight one connecting as many points as possible.

Observations

Part A

DATA TABLE 1

Description of Object	Material (if Known)	Mass (g)	Volume (cm^3)	Density (g/cm^3)
Bar	Iron Copper Aluminum			
Cube	Iron Copper Aluminum			

Name ______________________ Class ____________ Date ____________

Part B

DATA TABLE 2

Clay	Mass (g)	Volume (cm^3)
Sample 1		
Sample 2		
Sample 3		
Sample 4		
Sample 5		

Analysis and Conclusions

1. How do the densities of your two samples compare? Explain your observation.

____

____

____

____

2. What is the volume of a solid whose dimensions are 1.0 cm × 6.0 cm × 2.0 cm? Remember to include the proper units. ______________________________

____

3. If the mass of the object in question 2 is 60 g, what is its density?

____

____

4. Using your graph, determine the volume of a sample of clay that has a mass of 160 g.

____

5. Using your graph, determine the density of a piece of clay that has a volume of 3 cm^3.

____ g/cm^3

6. Using your graph (or data from your data table), determine the density of a piece of clay that has a mass of 25 g. ______________________________ g/cm^3

GRAPH

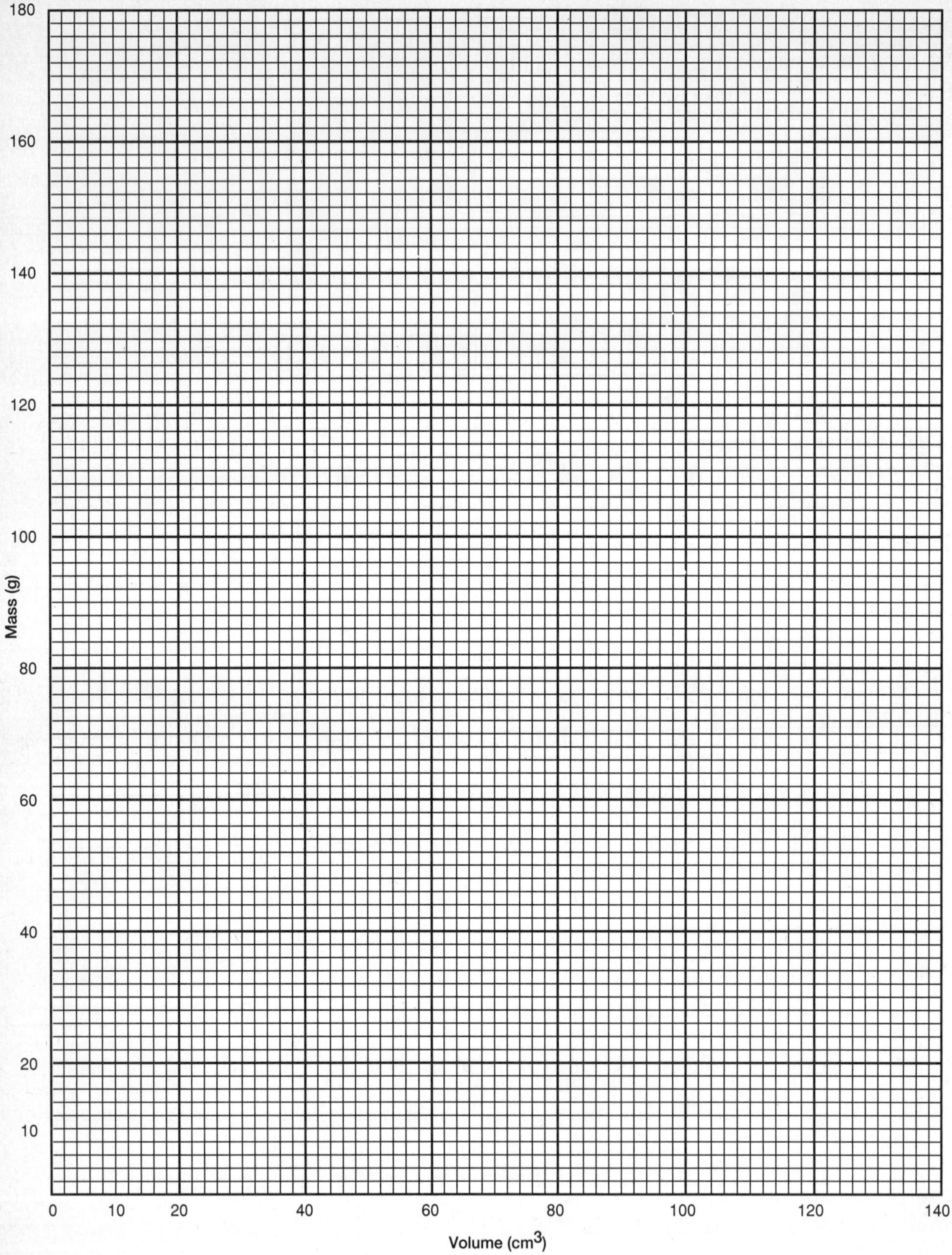

180
160
140
120
100
Mass (g)
80
60
40
20
10
0
10
20
40
60
80
100
120
140
Volume (cm3)

Name ______________________ Class ____________ Date ________

Critical Thinking and Application

1. Could the water displacement method be used to determine the volume of a rectangular solid as well as an irregular solid? ______________ Explain. ______________

____

____

2. If an object with a density of 5g/cm^3 is cut into two equal pieces, what is the density of each piece?

____

3. The diagrams below represent three samples of the same substance, each having a different size and shape. Arrange the letters of the samples to show the order by volume from largest to smallest. ______________________________

What is the density of A? B? C?

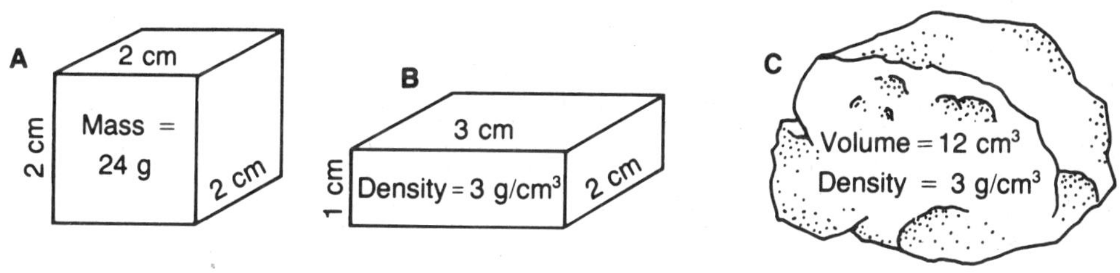

4. Explain how the results of this laboratory investigation show that differences in size and shape do not affect the density of a given substance.

____

____

____

5. Why is density such an important physical property?

____

____

____

____

____

Going Further

1. The density of water is 1 g/cm^3. An object will float in water if its density is less than 1 g/cm^3. If its density is greater than 1 g/cm^3, the object will sink. Given the following substances and their densities, determine whether each substance will float or sink in water.

aluminum	2.7 g/cm^3	___	gold	19.3 g/cm^3	___	chlorine	3.2 g/cm^3	___
arsenic	5.7 g/cm^3	___	neon	0.89 g/cm^3	___	uranium	19.0 g/cm^3	___
helium	0.18 g/cm^3	___	lithium	0.53 g/cm^3	___	potassium	0.86 g/cm^3	___

2. Substance X has a volume of 50 cm^3 and a mass of 160 g. Will substance X float or sink?

____

Substance Y has a volume of 140 cm^3 and a mass of 112 g. Will substance Y float or sink?

____

Name ______________________ Class ____________ Date ____________

Laboratory Investigation

General Properties of Matter **2**

Determining the Density of Liquids

Background Information

In the text you learned that mass and volume are general properties of all matter. Density is the ratio of mass to volume. The particular density of a specific kind of matter helps to identify it and distinguish it from other kinds of matter. Liquids have density, and it is possible to determine their densities in grams per milliliter (g/mL).

In this investigation you will determine the density of several liquids by measuring their mass and volume. You will also learn how to calculate the density of liquids by using a wood float.

Problem

How can you determine the density of liquids?

Materials *(per group)*

4 100-mL graduated cylinders
wooden cylinder about 10 cm long (A small pencil or thin dowel works well.)
laboratory balance
50 mL ethyl alcohol
oil (salad type)
salt water (several samples of different concentrations—A, B, C)
wax marking pencil

Procedure

1. Place the empty graduated cylinder on the laboratory balance. Record the mass in the Data Table.
2. Pour 50 mL of water into the graduated cylinder. Find the mass of the graduated cylinder and the water. Record this mass in the Data Table. Calculate the mass of the water by subtracting the mass of the empty graduated cylinder from the mass of the graduated cylinder with water. Record your answer in the Data Table. Keep the water in the cylinder for use in step 5.
3. Calculate the density of water by dividing the mass of the water by the volume.

$$\text{Density} = \frac{\text{Mass}}{\text{Volume}}$$

4. Using three other graduated cylinders, repeat steps 2 and 3 first using alcohol, then one of the samples of salt water, and then oil. Keep the liquids in the cylinders for use in step 5.

5. Carefully place the wooden cylinder in each liquid in the graduated cylinders, one at a time. See Figure 1. Note the level of the wooden cylinder in the water, the alcohol, the salt water, and the oil. In which liquid does the wooden cylinder float highest? In which liquid does the wooden cylinder float lowest?

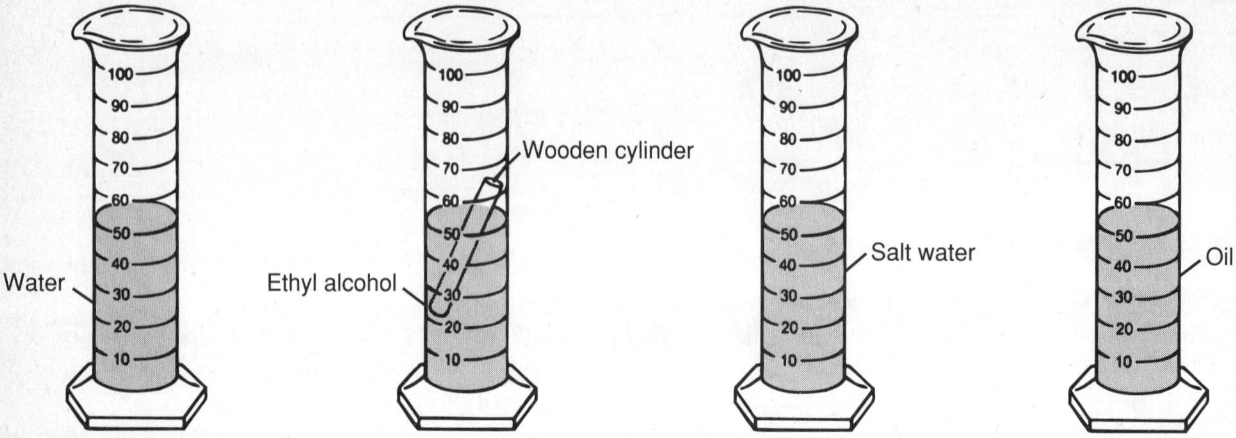

Figure 1

6. Use the wooden cylinder to test the remaining samples of salt water. With a wax marking pencil, mark the float levels of the salt water on your wooden cylinder.

Observations

DATA TABLE

Liquid	Mass of Empty Graduate (g)	Mass of Graduate and Liquid (g)	Mass of Liquid (g)	Volume of Liquid (mL)	Density of Liquid (g/mL)
Water				50	
Alcohol				50	
Salt water				50	
Oil				50	

Analysis and Conclusions

1. Which sample of salt water was the most dense? How did you know?

____

____

____

2. List the four liquids you used in this experiment in order of increasing density.

____

____

Name ______________________ Class ____________ Date ____________

Critical Thinking and Application

1. Which has the greater mass, 1 L of water or 1 L of alcohol? Why?

____

2. Which takes up a greater volume, 1000 g of water or 1000 g of alcohol? Why?

____

____

____

3. Which is more dense, 1 mL of water or 50 L of water? Why?

____

____

4. Predict what would happen if all the liquids used in this lab were poured into one beaker.

____

____

____

____

Going Further

Determine the density of each sample of salt water used in this investigation. Describe the method you used. Construct a table of your results.

Name ______________________________ Class ______________ Date ____________

Laboratory Investigation

General Properties of Matter **3**

Determining Specific Gravity

Background Information

One of the physical properties of matter that may be used to identify it is its specific gravity. Specific gravity is a comparison between the density of a substance and the density of pure water. Since the density of pure water is 1.0 g/mL, a substance with a specific gravity of 3.5, for example, would be 3.5 times denser than water.

In this investigation you will determine the specific gravity of several different substances by finding the mass of each and comparing it to the mass of the same volume of water.

Problem

How can an object's specific gravity be determined?

Materials *(per student or per group)*

mineral sample
rock sample
rubber stopper
cork
triple-beam balance
overflow can
250-mL beaker

Procedure

1. Find the mass of the empty, dry beaker. Record the mass in the Data Table.
2. Find the mass of the first object. Record the mass in the Data Table.
3. Fill the overflow can with water so that the water is above the spout. Allow the excess water to spill out into the sink.
4. Carefully drop the object into the overflow can and catch the displaced water in the beaker. If the object floats, push it down with a pencil or pen so that it is completely under water.
5. Find the mass of the beaker plus the water. Calculate the mass of the water alone and record it in the Data Table.
6. Calculate the specific gravity of the object according to the following ratio:

$$\text{Specific gravity} = \frac{\text{Mass of object}}{\text{Mass of displaced water}}$$

7. Repeat steps 1 through 6 for each additional object.

Observations

DATA TABLE

Object	Mass of Beaker	Mass of Water and Beaker	Mass of Displaced Water	Mass of Object	Specific Gravity

Analysis and Conclusions

1. When the object is placed in the overflow can, it displaces a certain volume of water. How does the volume of displaced water compare to the volume of the object?

____

____

2. What is the relationship between specific gravity and density?

____

____

____

3. If the object floats in water, why is it necessary to submerge it to obtain accurate results? ____

____

____

____

Critical Thinking and Application

1. Explain why the specific gravity of certain objects, such as table salt, cannot be determined by the method used in this investigation.

____

____

____

____

Name ______________________ Class ______________ Date ______________

2. If two objects have the same specific gravity, does it mean that they are made of the same kind of matter? Explain. ______________________

____

____

____

3. How could you use specific gravity to help you find out if a piece of gold jewelry is made of real gold or some other substance? ______________________

____

____

____

____

____

____

Going Further

This is not the only method of finding specific gravity. Use reference books to find out about other methods that may be used to determine specific gravity.

Name ______________________ Class ______________ Date ____________

Laboratory Investigation

Physical and Chemical Changes **4**

Investigating Phase Changes

Background Information

Pure matter can exist in three phases: solid, liquid, and gas. A phase change occurs when matter changes from one phase to another. Phase changes occur when heat is added or removed from a substance. The melting point of a substance is the temperature at which the solid-liquid phase change takes place. The freezing point is the temperature at which the liquid-solid phase change takes place. Notice that the melting point of a substance is the same as its freezing point. At this temperature, solid and liquid can exist together.

In Part A of this investigation you will observe what happens when the compound paradichlorobenzene (PDB) is cooled from a liquid to a solid, or freezes. In Part B the procedure will be reversed, and the solid will be heated until the substance melts and reaches a temperature above its melting point. The data obtained from Parts A and B will be used to construct a graph.

Problem

What temperature changes occur when a substance changes phase?

Materials *(per group)*

4 g paradichlorobenzene crystals
test tube
Celsius thermometer
hot plate
test-tube holder or clamp
2 400-mL beakers
safety goggles
heat-resistant gloves
test-tube rack
colored pencils
glass-marking pencil

Procedure
Part A

1. Half fill the two beakers with water. Heat one of the beakers on the hot plate until the temperature of the water is about 90°C. Do not boil the water.

2. Obtain a test tube of PDB crystals and place it in the beaker of hot water. Allow the PDB to melt completely. As soon as it is melted, insert a thermometer in it and record the temperature. See Figure 1.

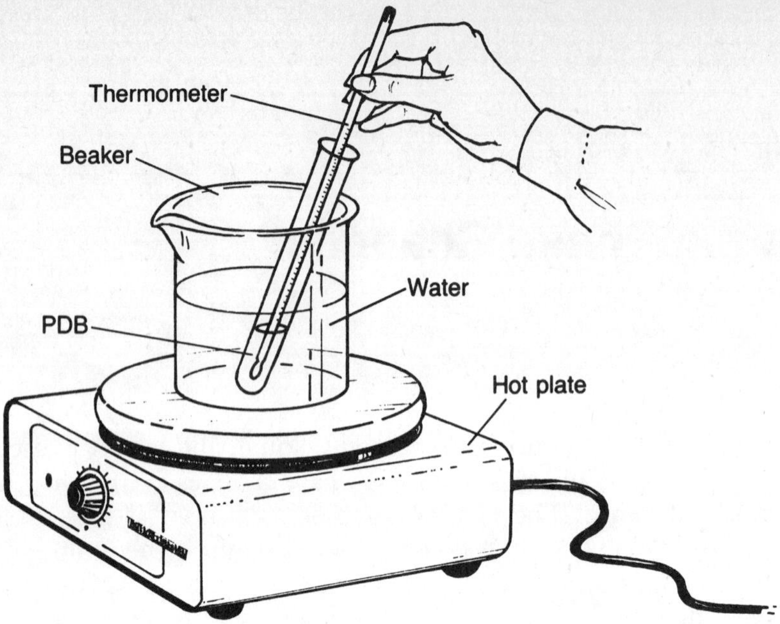

Figure 1

3. Remove the test tube from the beaker and place it in the beaker of cool water. Hold the thermometer in the PDB so that it does not touch the sides of the test tube, and record the temperature every 30 seconds in Data Table 1. Between readings, stir the PDB constantly until it becomes solid. Be sure to note in your data when the liquid PDB first starts to solidify and when there is no liquid PDB left. Continue to record the temperature until it drops to 40°C.

4. Note what happens to the volume of liquid PDB as it turns into solid PDB.

Part B

5. Be sure the beaker of water on the hot plate is still heated as in step 1. Place the test tube with solid PDB in the beaker of hot water. Record the temperature of PDB in Data Table 2 as soon as you place it in the hot water. Read and record the temperature every 30 seconds until solid PDB has changed completely to liquid.

6. When the thermometer is able to move, use it to stir the mixture of solid-liquid PDB. Continue stirring and recording the temperature every 30 seconds until the temperature of the sample reaches 60°C. Note in your data when the solid PDB begins to melt and when it has completely melted.

7. Remove the thermometer and wipe it clean. Remove the test tube from the hot water and allow it to cool in the test-tube rack before replacing the cork and returning it to your teacher.

8. Graph your data on Graph 1. Use one color for the data you obtained while cooling the PDB and another color for the data you obtained while heating the PDB. Provide a key to indicate what each color represents.

Name ______________________ Class ______________ Date ____________

Observations

DATA TABLE 1 (Cooling Curve)

Time (min)	Temperature (°C)	Time (min)	Temperature (°C)
½		7½	
1		8	
1½		8½	
2		9	
2½		9½	
3		10	
3½		10½	
4		11	
4½		11½	
5		12	
5½		12½	
6		13	
6½		13½	
7		14	

DATA TABLE 2 (Heating Curve)

Time (min)	Temperature (°C)	Time (min)	Temperature (°C)
½		7½	
1		8	
1½		8½	
2		9	
2½		9½	
3		10	
3½		10½	
4		11	
4½		11½	
5		12	
5½		12½	
6		13	
6½		13½	
7		14	

GRAPH 1

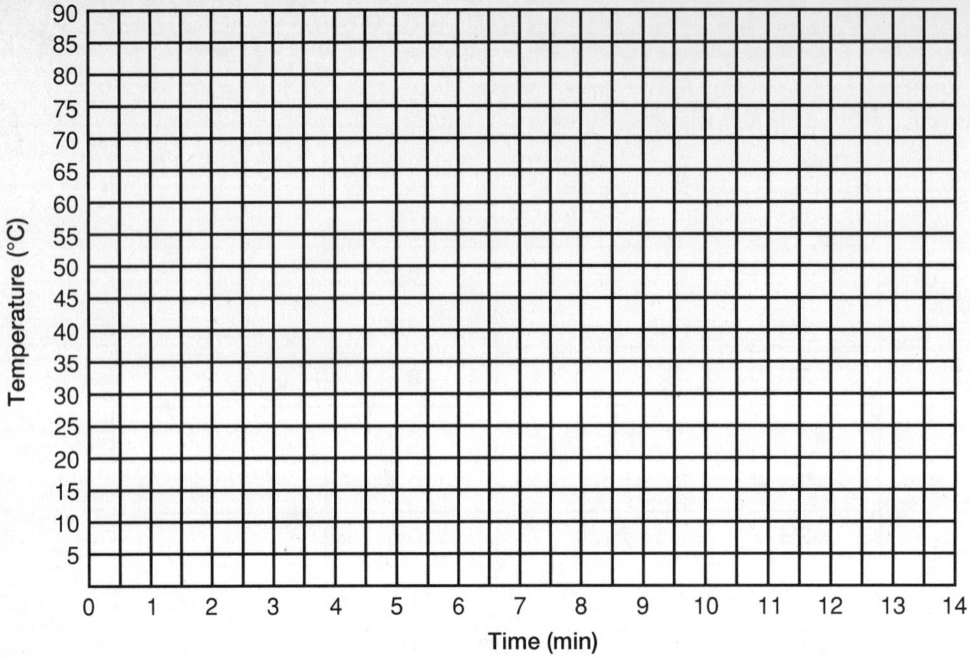

Analysis and Conclusions

1. At what temperature does PDB become solid? ______ How does this compare to the temperature at which solid PDB become a liquid? ____________________

____

2. Looking at the slopes of the lines on your graph, how many different sections make up your graph? ____________________

3. Describe what occurs in the section on your graph where the line is horizontal.

____

____

____

____

4. What happened to the volume of PDB as it became a solid?

____

What does this indicate about the density of solid PDB as compared to liquid PDB?

____

____

____

Name ______________________ Class ____________ Date ____________

Critical Thinking and Application

1. As the PDB was melting, heat was being added to it but the temperature did not rise. Where did the heat energy go? ______________________

2. If you used ice instead of PDB in this experiment, how do you think your results would be different? ______________________

3. If you used twice as much PDB, how would your results be different? Explain.

4. A student wanted to solve the problem: "Which freezes fastest, hot water or cold water?" Using the scientific method, describe an experiment the student might do to find the answer. ______________________

What would your hypothesis be if you were doing this experiment?

Going Further

Do the experiment you have described in answer to question 4. Record your observations and conclusions.

Physical and Chemical Changes

Background Information

Matter is constantly changing. The two kinds of changes that occur in matter are physical and chemical changes. In a physical change, no new substances are formed. However, physical properties such as size, shape, color, or phase may change. Dissolving, melting, evaporating, and grinding are examples of physical change.

As a result of chemical change, one or more "new" substances with new and different properties are formed. The new substances are different from the original substance. Burning and the rusting of iron are examples of chemical change.

In this investigation you will observe physical and chemical changes and learn to recognize each type of change when it occurs.

Problem

What are the differences between physical and chemical changes?

Materials *(per student)*

birthday candle
aluminum foil (15 cm × 15 cm)
modeling clay
small piece of paper
watch glass
Bunsen burner
scoop
magnesium ribbon (1 cm long)
1 M hydrochloric acid
safety goggles
2 test tubes
test-tube rack
test-tube clamp
insulating pad
matches
table salt
dropper bottle of 0.1 M silver nitrate

Procedure

1. Take a small piece of modeling clay and place it on the square of aluminum foil. Firmly place a candle in the clay so that it is well supported. Light the candle and allow it to burn while you continue with the rest of the investigation. Record your observations of the burning candle in the space provided in Observations.

2. Tear the piece of paper into small pieces and place them on the watch glass. Place the watch glass and pieces of paper on the insulating pad. Light the pieces of paper with a match and allow them to burn completely. Record your observations of the burning paper.

3. Add a small scoop of table salt to a test tube that has been half-filled with tap water. Place your thumb over the top of the test tube and shake to dissolve the salt. Record your observations. Using the dropper, add 5 drops of silver nitrate to the salt water. Record your observations.

4. Place a small piece of magnesium ribbon in a test tube. Add 5 drops of hydrochloric acid to the test tube. Touch the bottom of the test tube with your fingertips. Record your observations.

Observations

1. What did you observe as the candle burned? ______________________________

____

____

What was left after the candle burned? ______________________________

____

2. What did you observe as the paper burned? ______________________________

____

What was left after the paper burned? ______________________________

____

3. What did you observe when you added the salt to the water in the test tube and shook it? ______________________________

____

What did you observe when the silver nitrate was added to the salt water?

____

____

4. What did you observe when the hydrochloric acid was added to the magnesium metal?

____

____

____

____

Name ______________________ Class ____________ Date ____________

Analysis and Conclusions

1. Identify each of the following as either a physical change or chemical change. Give a reason for your answer.

   a. Melting candle wax ______________________

   ____

   b. Burning a candle ______________________

   ____

   c. Tearing paper ______________________

   ____

   d. Burning paper ______________________

   ____

   e. Dissolving table salt ______________________

   ____

   f. Mixing salt water and silver nitrate ______________________

   ____

   g. Cutting a piece of magnesium ribbon ______________________

   ____

   h. Adding hydrochloric acid to magnesium metal ______________________

   ____

2. Describe two observations you might make when a physical change occurs.

   ____

   ____

   ____

3. Describe two observations you might make when a chemical change occurs.

   ____

   ____

   ____

Critical Thinking and Application

1. How could you show that dissolving the salt in water resulted in a physical change?

2. How could you show that adding acid to the magnesium ribbon resulted in a chemical change?

3. The following changes can sometimes indicate that a chemical change has occurred. Explain how each change might result from a physical, not a chemical, change.

   a. Change of color

   b. Loss of mass

   c. The substance seems to "disappear."

Going Further

Write out a recipe that involves cooking or baking. Identify each step in the recipe as resulting in either a physical change or a chemical change in the ingredients.

Laboratory Investigation

Mixtures, Elements, and Compounds **6**

Determining Solubility

Background Information

Several factors affect the rate at which solids dissolve in liquids. These factors include the nature of the solute and solvent, the temperature, and the degree of fineness to which the solute has been ground. There is no general rule to predict how much solute will dissolve in a given solvent, but you can determine the effect certain variables have on the rate of solution.

In this investigation you will determine the amount of potassium nitrate that can be dissolved in 10 mL of water at a given temperature.

Problem

How can you determine the solubility of a substance in water?

Materials *(per group)*

25 mL potassium nitrate
water
100-mL graduated cylinder
250-mL beaker
Celsius thermometer
Bunsen burner
2 test tubes
laboratory balance
1 test-tube holder
ice
glass stirring rod
safety goggles

Procedure

1. Place a small sheet of plain paper on the laboratory balance. Record the mass of the paper. Adjust the balance so that it registers 25 g more than the mass of the paper alone.
2. Slowly and carefully add potassium nitrate to the paper until the balance is again level. In this way, you have poured out 25 g of potassium nitrate.
3. Pour 10 mL of water into one test tube. Warm the test tube over the Bunsen burner until the first temperature assigned to you by your teacher is reached. Try to maintain this temperature throughout the next step.
4. Pour a small amount of potassium nitrate into the test tube. Stir carefully. If the potassium nitrate dissolves completely, add a little more. Continue until no more dissolves and a few small grains settle to the bottom of the test tube.
5. Record the exact temperature of the solution.

6. Find the mass of the paper and the remaining potassium nitrate again. Determine the amount of potassium nitrate you used by subtracting this amount from the original mass.
7. Repeat the procedure with another 10 mL of water. However, this time cool the water in ice until the second temperature assigned to you by your teacher is reached. Again determine the amount of potassium nitrate that dissolves in the water. Record the exact temperature.
8. Report your information to your teacher. He or she will compile all the information obtained by the class. In this way, you will find out how much potassium nitrate dissolves in 10 mL of water over a wide temperature range.
9. Graph your results and the results of your classmates in Graph 1.

Observations

GRAPH 1

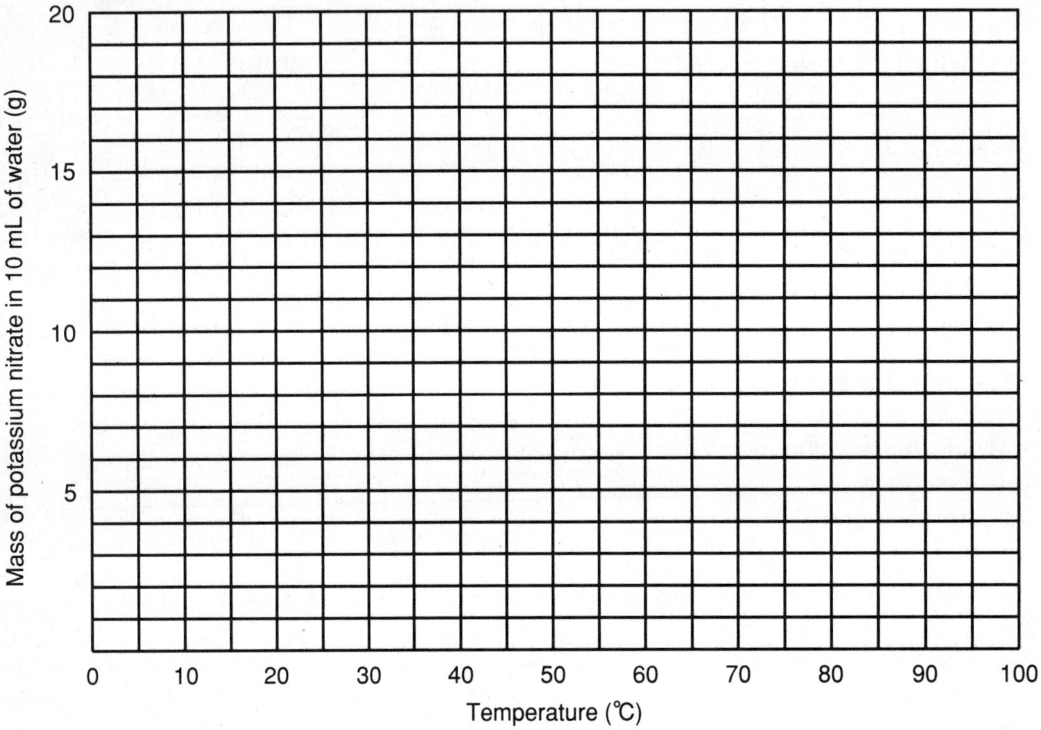

Analysis and Conclusions

1. What effect does temperature have on the amount of potassium nitrate that can be dissolved in a given amount of water? ____

____

____

2. From the graph, predict how much potassium nitrate would dissolve in 10 mL of water at 60°C. ____

3. How much potassium nitrate do you think would dissolve in 100 mL of water at 60°C?

____

4. To what temperature would 10 mL of water have to be heated to completely dissolve 14 g of potassium nitrate? ____

Critical Thinking and Application

1. If the temperature of a saturated solution of potassium nitrate were to drop, what would you notice? ____

____

2. If 10 mL of a saturated solution of potassium nitrate were to cool from 60°C to 10°C, how much potassium nitrate would be found on the bottom of the test tube?

____

3. Suppose you had measured the solubility of KNO_3 only at 10°C and 90°C. How might your solubility graph be inaccurate? ____

____

____

____

4. Based on the graph, how much KNO_3 do you think would dissolve in 10 mL of water at 100°C? ____

Going Further

Determine the solubility graph for sodium chloride (table salt). How does this graph differ from the graph for potassium nitrate?

Name ______________________ Class ____________ Date ____________

Laboratory Investigation

Mixtures, Elements, and Compounds **7**

Elements, Mixtures, and Compounds

Background Information

When two elements are combined, a chemical reaction does not always occur between them. Instead, the two elements form a mixture in which the properties of the two elements are retained.

When a chemical reaction does take place, the two elements form a compound. The compound has physical and chemical properties that the individual elements alone do not have.

The two elements in a mixture can be separated on the basis of differences in their physical properties. The two elements in a compound, however, can only be separated as a result of another chemical reaction.

In this investigation you will observe how physical and chemical properties are affected when two elements are combined.

Problem

How do the properties of elements differ in mixtures and in compounds?

Materials *(per group)*

triple-beam balance
forceps
scoop
Bunsen burner
matches
sulfur powder
iron powder
hand lens
test tube
test-tube clamp
250-mL beaker
magnet
safety goggles

Procedure

1. On your triple-beam balance, measure out 3 g of sulfur powder on a piece of paper.
2. Using the hand lens, observe the color of the sulfur and the size and shape of the particles. Record your observations in the Data Table.
3. Lift the edge of the paper containing the sulfur, place the magnet underneath, and move the magnet to see if it has any effect on the sulfur. Record your observations in the Data Table.
4. On your triple-beam balance, measure out 5 g of iron powder on a piece of paper.
5. Using the hand lens, observe the color of the iron powder and the size and shape of the particles. Record your observations in the Data Table.
6. Lift the edge of the paper containing the iron, place the magnet underneath, and move the magnet to see if it has any effect on the iron. Record your observations in the Data Table.

7. Mix the iron and sulfur together on one piece of paper. Use the scoop to make sure the mixture is uniform throughout. Repeat step 2 for the mixture. Record your observations in the Data Table.

8. Repeat step 3 for the mixture. Record your observations in the Data Table.

9. Half-fill the beaker with cold water and set it aside. Carefully place the iron-sulfur mixture in a test tube. Light the Bunsen burner. Holding the test tube with the test-tube clamp, heat the iron-sulfur mixture for about 5 minutes. **CAUTION:** *Make sure that the mouth of the test tube is pointed away from you and others.* Move the test tube around in the flame so that all of the contents are evenly heated. Record your observations in the Data Table.

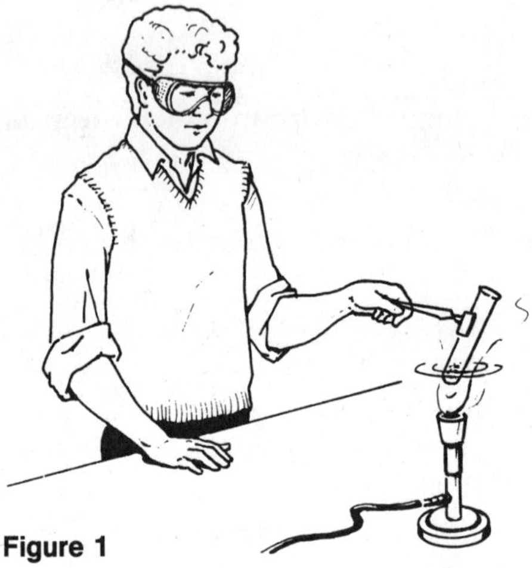

Figure 1

10. When you see no further changes occurring in the test tube, shut off the Bunsen burner and immediately put the test tube into the beaker of cold water. If the test tube does not break, wrap it in several layers of paper towel and hit it with a hammer to crack the glass.

11. Using the forceps, hand lens, and magnet, examine the substance formed in the test tube. Record your observations in the Data Table.

Observations

DATA TABLE

Physical Properties	Sulfur	Iron	Iron-Sulfur Before Heating	Iron-Sulfur After Heating
Color				
Shape of particle				
Size of particle				
Effect of magnet				

What did you observe when you began heating the test tube containing the mixture of sulfur and iron? ______________________________

____

____

Name ______________________ Class ____________ Date ____________

Analysis and Conclusions

1. How did the properties of sulfur alone compare with the sulfur in the unheated iron-sulfur combination? ______________________

____

2. How did the properties of iron alone compare with the iron in the unheated iron-sulfur combination? ______________________

____

3. How did the properties of sulfur alone compare with the iron-sulfur combination after it was heated? ______________________

____

4. How did the properties of iron alone compare with the iron-sulfur combination after it was heated? ______________________

____

____

5. What kind of substance was the iron-sulfur combination before heating?

____

6. What was the effect of heating the iron-sulfur combination?

____

7. What kind of substance was the iron-sulfur combination after heating?

____

Critical Thinking and Application

1. Explain how you know that mixing the iron and sulfur together on the paper results in a mixture of the two substances. ______________________

____

____

____

2. Explain how you know that heating the iron and sulfur together in the test tube results in a new compound composed of chemically combined iron and sulfur.

3. Why was it necessary to heat the test tube in order for a chemical reaction to occur?

4. The element sodium reacts explosively with water and the element chlorine is a yellowish-green poisonous gas. When chemically combined, sodium and chlorine form nonexplosive, nonpoisonous table salt. Explain how this can be.

5. Identify each of the following as an element, a mixture, or a compound.

   a. aluminum foil ______

   b. air ______

   c. water ______

   d. salt water ______

   e. copper wire ______

   f. steel ______

Going Further

The chemical combination of iron and sulfur produces the compound iron sulfide, also known as the mineral pyrite. Consult reference sources to find the properties and uses of pyrite.

Making Predictions Using Indirect Evidence

Background Information

How do scientists know what an atom looks like? After all, it is impossible to see an atom. Yet scientists have constructed models of an atom based on observations of how atoms behave. These observations are considered to be one type of indirect evidence for the structure of an atom.

Another type of indirect evidence comes from making predictions. Based on the model they have developed, scientists predict how atoms will behave in certain circumstances. They design experiments to test these predictions. If the predictions are shown to be accurate, they are taken as additional indirect evidence that the model is correct.

In this investigation you will use indirect evidence to determine the properties of an object you cannot see.

Problem

How can you determine the characteristics of something you cannot observe directly?

Materials *(per pair of students)*

shoe box, with cover, that contains small objects
laboratory balance

Procedure

1. Obtain a sealed shoe box from your instructor.
2. Without opening the box, perform several tests such as tipping it, shaking it, sliding it, and finding its mass. In each case, record your observations in the Data Table. In each case, the observation could be of a sound or of the way the object behaved (rolling, sliding, and so on). You might also wish to find the mass of an identical box and lid that is empty. The difference in mass would be the mass of the object or objects inside.
3. On the basis of your observations, sketch the object or objects in Drawing 1. Include any other characteristics, such as mass.

Observations

Describe some of the ways you tested the unknown objects to determine their characteristics. Record your descriptions in the Data Table.

DATA TABLE

Test	Observations

Analysis and Conclusions

DRAWING 1

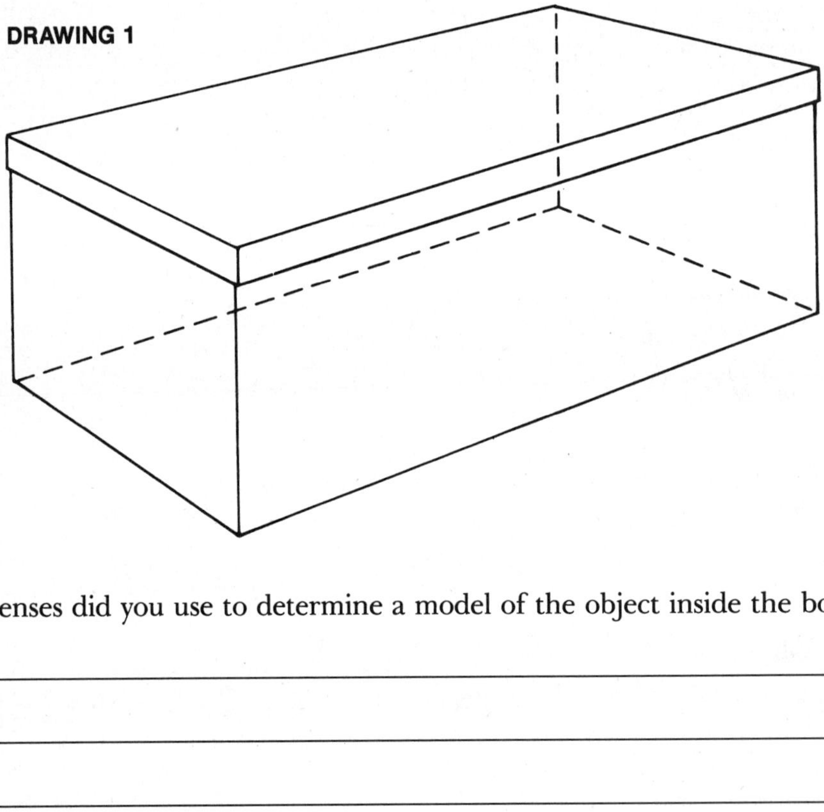

1. What senses did you use to determine a model of the object inside the box?

____

____

____

2. How does this kind of activity compare with the way scientists have learned about the atom? _______________

____

____

____

Name ______________________ Class ____________ Date __________

Critical Thinking and Application

1. In what ways could you test your model of the contents of the box?

2. What kinds of special instruments could also be used to make indirect observations of the contents of the box?

3. If you used some of the special instruments you mentioned and the results did not agree with your previous determination of the contents of the box, what would you need to do?

4. Open the box and directly observe the object(s) inside. How does your observation compare with your indirect determination?

   What does this tell you about the current model?

Going Further

1. Glue an object inside a shoe box. Trade boxes with another student who has done the same. Using knitting needles, determine the location and shape of the object by sticking the needles through the box.
2. How would you determine what kind of liquid was inside of a full, closed, opaque bottle? How would it be helpful if you had an identical empty bottle?

Name ______________________ Class ______________ Date ______________

Laboratory Investigation

Atoms: Building Blocks of Matter **9**

Investigating Rutherford's Model of the Atom

Background Information

In 1911, Ernest Rutherford discovered that most of the atom is empty space. Almost all of the atom's mass is concentrated in an extremely small volume called the nucleus. The nucleus is located in the center of the atom. Rutherford discovered the existence of the nucleus in his famous scattering experiment. In this experiment, he aimed tiny, positively charged alpha particles at a very thin piece of gold foil. Rutherford observed that almost all the positively charged alpha particles passed through the gold foil with almost no deflection, or bending. This could only mean that the gold atoms in the sheet were mostly empty space!

However, Rutherford also observed that some particles were greatly deflected. In fact, a few bounced almost straight back from the gold foil. What could this mean? Rutherford knew that positive charges repel other positive charges. So he proposed that an atom had a small, positively charged center. He called this center the nucleus. In this investigation you will duplicate Rutherford's experiment.

Problem

How can the presence of an "invisible" charged particle be detected?

Materials *(per group)*

sheet of white paper,
  50 cm × 20 cm
pencil
metric ruler
target platform with "nucleus"
  and ramp attached
tape
sheet of colored paper,
  30 cm × 20 cm
2 strips of posterboard,
  30 cm × 2 cm
strip of posterboard,
  20 cm × 2 cm
"alpha particle"
flour

Procedure

Part A Constructing Your Setup

1. On the sheet of white paper, draw a series of lines lengthwise, 2 cm from each other. When you are finished there should be about ten parallel lines.
2. Tape this paper to the elevated ramp, as shown in Figure 1.

Figure 1

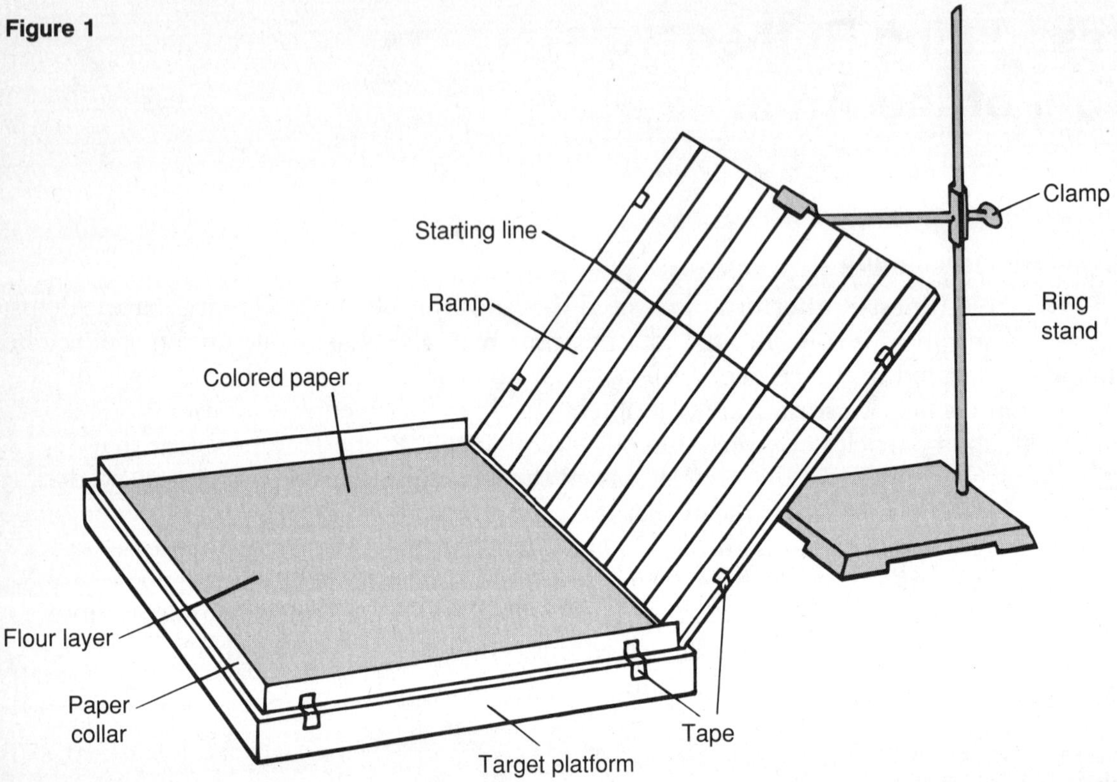

3. Tape the piece of colored paper to the target platform. **Note:** *Be sure that the edge of the paper does not interfere with the sliding of the "alpha particle" down the ramp to the platform.*
4. Place the alpha particle about halfway up the ramp. Make sure that the smooth surface of the alpha particle is facing down.
5. Allow the alpha particle to slide down the ramp and cross the platform. **Note:** *The alpha particle should not come to a stop until it is at least halfway across the target platform.* If it stops too soon, release it from a higher position on the ramp.
6. Adjust the height of the ramp until the alpha particle slides about 20 cm across the target platform before stopping. Using the ruler, draw a line across the ramp at this height.
7. Carefully sprinkle some flour on the platform so that the platform is entirely covered with flour. Use your ruler to create a smooth and even layer of flour. The flour will aid you in observing the path of the alpha particle.

Part B Duplicating Rutherford's Experiment

1. Release the alpha particle on the line on the ramp closest to you. Make sure that you position the alpha particle at your starting line.
2. Using the pencil, draw the exact path the alpha particle took in the flour on the colored paper. Label this line 1.

3. Smooth out the surface of the flour with your ruler.
4. Release the alpha particle along the next line. Draw its path on the colored paper. Label this line 2.
5. Repeat steps 2 and 3 until the alpha particle has been released from each of the vertical lines.
6. When you are finished, carefully remove the colored paper from the target platform. Carefully dispose of the flour.

Observations

1. On which areas of the target platform were the paths of the alpha particle straight lines? ______________________

2. How many lines were deflected, or bent, on the target platform?

____

3. From your observations, locate the target "nucleus" on your colored paper. Draw its approximate size and shape. Attach the colored paper to this laboratory investigation.

Analysis and Conclusions

1. If the target platform represents an atom, is the target nucleus the same size as the atom? Explain. ______________________

____

2. Did the target nucleus attract or repel the alpha particle? ______________

Explain. ______________________

____

Critical Thinking and Application

1. In Rutherford's actual experiment, electric forces deflected the positive alpha particles from the positive target nuclei. In your duplication of Rutherford's experiment, what type of force did you use to deflect the alpha particle? How could you prove that you used this type of force? ______________________

____

____

____

2. In Rutherford's actual experiments, nearly all the alpha particles passed straight through and were not deflected at all. What does this tell you about the size of the nucleus of an atom? ______________________________

____

As a result of these experiments, of what does an atom seem to mostly consist?

____

____

3. In other experiments, Rutherford aimed beta particles, which are high-speed, negatively charged electrons, at the foil and screen. Predict the behavior of the beta particles. ________

____

____

____

4. Rutherford repeated these experiments using foils made of metals other than gold. Predict the results of these experiments. ______________________________

____

____

Going Further

Find out about the contributions made by Crookes, Thomson, and Millikan to the discovery of the electron.

Name ______________________ Class ____________ Date ____________

Laboratory Investigation

Atoms: Building Blocks of Matter **10**

Relating Electrons and Probability

Background Information

Early theories of the structure of the atom described the movement of electrons around the nucleus as similar to the movement of the planets around the sun. Today scientists know this is not the case. Electrons do not travel around the nucleus in fixed orbits. Electrons move in an area known as the electron cloud. The electron cloud is a region in which electrons are likely to be found. Within the electron cloud, electrons are arranged in energy levels. Energy levels represent the most probable location in which an electron can be found. An energy level should not be confused with a specific path. For electrons do not have a path. In fact, scientists can speak only of the chances, or probability, of finding electrons at various locations—not of their exact position. In this investigation you will get a better understanding of probability and how it relates to electrons.

Problem

How can the movement of electrons outside the nucleus be described?

Materials *(per group)*

1 die
masking tape
1 sheet of graph paper having
1½ squares to the centimeter

Procedure

1. Prepare your die in the following way: Tape all six sides. Mark three of the sides with one dot each, two of the sides with two dots each, and the remaining side with three dots.
2. Select a square near the center of the graph paper and color it red. This red square will represent the nucleus.
3. Roll the die and with each roll pencil in a square according to the following rules.
   - If a one is rolled, pencil in any square that is between 0 and 3 cm from the nucleus.
   - If a two is rolled, pencil in any square that is between 3 and 5 cm from the nucleus.
   - If a three is rolled, pencil in any square that is between 5 and 7 cm from the nucleus.
4. Repeat this procedure of rolling the die and marking the graph for 50 throws. Record your results in the Data Table.

Observations

DATA TABLE

	Number of Squares Penciled In
0–3 cm (first energy level)	
3–5 cm (second energy level)	
5–7 cm (third energy level)	

1. The modern view of the movement of electrons in an atom is based on the concept of probability. What is the definition of probability? ______________________________

____

____

____

2. In which range do you have the most darkened squares on your diagram?

____

____

3. Compare your diagram to that of a classmate. Are they identical?

____

____

In what ways are they alike? Different? ______________________________

____

____

____

Analysis and Conclusions

1. Based on your data, in what energy level is an electron most likely to be found?

______________ Least likely to be found? ______________

Name ______________________ Class ____________ Date ____________

2. If each square that you penciled in represents a chance of finding an electron in a particular location around the nucleus, where would you look first for an electron?

Explain your answer. ______________________

____

____

3. Can the exact position of an electron around the nucleus be determined?

____

What can be known about electrons? ______________________

____

Critical Thinking and Application

1. Mathematically, the probability of an event occurring is equal to the number of favorable outcomes divided by the number of possible outcomes. According to the way you marked your die at the beginning of this investigation, what is the probability of an electron being found in the first energy level? The second? The third?

____

____

2. Suppose you had rolled the die 100 times. How do you think your results would have compared with the results you obtained by rolling the die 50 times?

____

____

____

____

3. Suppose you wanted to determine the probability of a student being found at various locations in your school building at 10 AM. How might you investigate this problem?

____

____

____

____

____

Going Further

Each energy level in an atom can hold only a certain number of electrons. The first energy level, closest to the nucleus, can hold at most 2 electrons. The second and third energy levels are considered full when they have 8 electrons each. Using this information, fill in the chart below.

Element	Number of Electrons	Electron Arrangement			
		First Energy Level	Second Energy Level	Third Energy Level	Fourth Energy Level
Hydrogen					
Helium					
Carbon					
Oxygen					
Neon					
Sodium					
Aluminum					
Sulfur					
Chlorine					
Calcium					

Name ________________ Class ________ Date ________

Laboratory Investigation

Classification of Elements: The Periodic Table **11**

The Alkaline Earth Elements

Background Information

The arrangement of the elements in the periodic table is one of the most important achievements in modern chemistry. The physical and chemical properties of elements change in a regular pattern as you go both across the rows and down the columns of the periodic table. As a result, when elements close to each other in a row or column are compared, they have many of the same properties. However, when elements farther away from each other in a row or column are compared, they have more dissimilar properties.

The elements in Group 2 of the periodic table are known as the alkaline earth elements. Like all members of a group, or family, of elements, they have certain properties that change in a regular pattern within the group. One of these properties is the ability to form a precipitate, or solid substance, as a result of a chemical reaction. The precipitate cannot dissolve in water and eventually settles to the bottom of the container.

In this investigation you will compare the abilities of the alkaline earth elements to form precipitates as a result of a chemical reaction.

Materials *(per group)*

safety goggles
spot plate
sheet of notebook paper
dropper bottles of:
- magnesium nitrate
- calcium nitrate
- strontium nitrate
- barium nitrate
- potassium carbonate
- potassium sulfate
- potassium chromate

Procedure

1. Place the spot plate in the center of a sheet of notebook paper so that there are 4 spots running down and 3 spots running across. See Figure 1.
2. Along the side of the notebook paper next to each of the four spots, write the names of the four alkaline earth elements that are present in each nitrate compound listed in the materials you are using. Write them in the same order in which they are listed. See Figure 1.

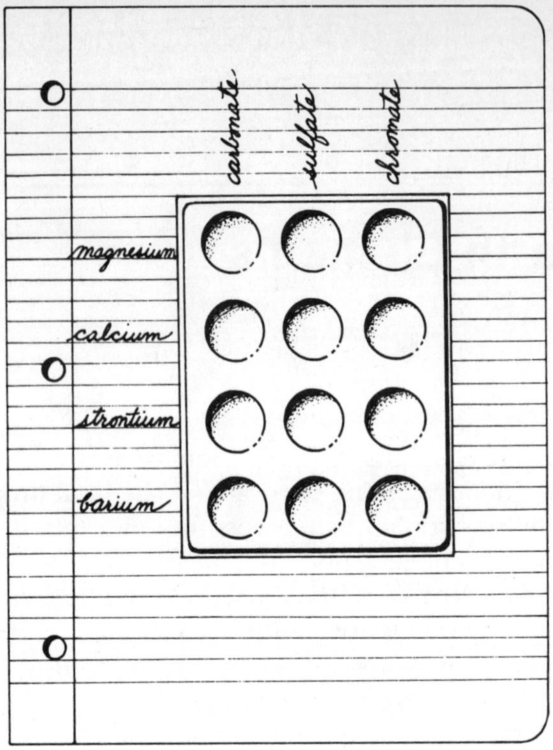

Figure 1

3. Along the top of the notebook paper next to each of the three spots, write the names of the three substances that are combined with potassium in the materials you are using. See Figure 1.

4. Put on your safety goggles. Place 3 drops of potassium carbonate in each of the four spots under the word Carbonate. Place 3 drops of potassium sulfate in each of the four spots under the word Sulfate. Place 3 drops of potassium chromate in each of the four spots under the word Chromate. Be very careful not to mix the liquid from one spot with the liquid from another.

5. Take the dropper bottle of magnesium nitrate and place 3 drops in each of the three spots in the row labeled Magnesium. Observe each spot carefully and record the result in the Data Table. Repeat this procedure using the dropper bottles containing calcium nitrate, strontium nitrate, and barium nitrate. Be very careful not to mix the liquid from one spot with the liquid from another.

6. After recording your results, wash your spot plate thoroughly with soapy water and a brush.

Observations

DATA TABLE

Alkaline Earth Metal	Carbonate	Sulfate	Chromate
Magnesium			
Calcium			
Strontium			
Barium			

Name ______________________ Class ____________ Date ____________

Analysis and Conclusions

1. Was there evidence of a chemical reaction occurring in any of the spots? ____________

   Explain your answer. ______________________

2. Which alkaline earth element formed the smallest number of precipitates? ____________

3. Which alkaline earth element formed the greatest number of precipitates? ____________

4. What is the relationship between the number of precipitates formed and the location of the alkaline earth element on the periodic table? ______________________

5. If the ability of an alkaline earth element to form a precipitate is an indication of its ability to chemically react with other substances, which is the most reactive element?

   ____

   The least reactive?

   ____

6. List the alkaline earth metals in order of their chemical reactivity, starting with the most reactive. ______________________

7. How does the order of the elements you listed in question 6 compare to their order in the periodic table? ______________________

Critical Thinking and Application

1. Why were you cautioned not to mix the solution in one spot with the solution in another spot on your spot plate? ______________________

2. Group 1 in the periodic table is known as the alkali metals. Based on your investigation of the Group 2 elements, predict the comparative reactivity of the elements in Group 1 of the periodic table.

---

3. If you had a solution containing a mixture of magnesium, strontium, and barium, how could you separate the three elements?
Hint: Review the information in the Data Table.

Going Further

Perform the procedure you described in your answer to question 3. Be sure to check the procedure with your teacher before starting. Describe your investigation in a laboratory report that includes title, problem, materials, procedure, observations, and conclusions.

Laboratory Investigation

Classification of Elements: The Periodic Table **12**

Investigating the Activity Series of the Metals

Background Information

One way to arrange the elements in the periodic table is by order of their activity. A metal is said to be more active than another metal if it will replace the less active metal from a solution of one of its compounds. This can be observed because the metal with the lower activity is "plated out" as pure metal on the piece of metal with the higher activity. By performing a series of experiments, you can develop a list of metals in order of their activity.

In this investigation you will test zinc, copper, and lead to see if they will replace each other in solutions of their compounds. Using your results, you can develop a short list in order of activity.

Problem

How can you determine the relative activities of zinc, copper, and lead?

Materials *(per pair of students)*

- solutions
  - copper sulfate
  - lead acetate
  - zinc sulfate
- 10-cm strips
  - zinc
  - lead
  - copper
- sandpaper
- 9 small test tubes
- test-tube rack
- safety goggles
- masking tape

Procedure

1. Put on safety goggles.
2. Arrange the nine test tubes in three groups of three each. Using small pieces of masking tape, label the first three test tubes Lead Acetate, the next three Zinc Sulfate, and the last three Copper Sulfate. See Figure 1.

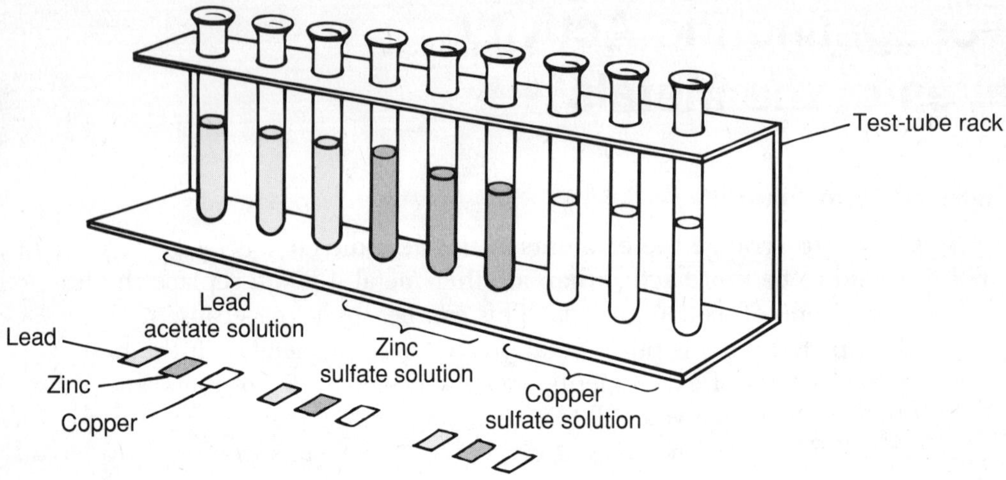

Figure 1

3. Fill the appropriately labeled three tubes one third full of lead acetate solution. Fill the appropriately labeled three tubes one third full of zinc sulfate solution. Fill the appropriately labeled remaining three tubes one third full of copper sulfate solution.
4. Clean the metal strips by rubbing with sandpaper. Using a separate tube for each strip of metal, place strips of lead, zinc, and copper in each of the three kinds of solutions.
5. After several minutes, examine the strips for evidence of a reaction; remove them from the solutions if necessary. Record your observations in the Data Table. Also note if the solution changed color.

Observations

DATA TABLE

	Lead Acetate	Zinc Sulfate	Copper Sulfate
Lead			
Zinc			
Copper			

1. Which metal did not react with any of the solutions it was placed in?

____

2. Which metal reacted with all but its own solution?

____

Name ______________________ Class __________ Date __________

Analysis and Conclusions

1. List the metals in order of decreasing activity. ______________________

____

2. How could you tell which of the three metals was the most active?

____

____

Critical Thinking and Application

1. Magnesium is more active than zinc. How would a strip of magnesium have reacted in the three solutions you studied? ______________________

____

2. Tin is less active than zinc but more active than lead. How would a strip of tin have reacted in this investigation? ______________________

____

3. A strip of silver does not react with any of the solutions tested. What must be true about silver? ______________________

____

4. What experiment could you perform in order to determine if iron is more active than copper? ______________________

____

____

5. Metals like platinum and gold are often said to be inactive. What does this mean?

____

____

Going Further

1. Read about plating with copper or silver. How is plating done commercially? How could you do it in the laboratory?
2. Determine the activity of aluminum relative to the other metals.